YOUR KNOWLEDGE HAS VALUE

AF167031

Clarias Gariepinus Fish as a Suitable Culture for Carrion Insects in Forensic Investigations using Dichlorvos as a Suicide Agent. Forensics and Crime

Emmanuel Tyokumbur

Bibliographic information published by the German National Library:

The German National Library lists this publication in the National Bibliography; detailed bibliographic data are available on the Internet at http://dnb.dnb.de.

ISBN: 9783346897428
This book is also available as an ebook.

© GRIN Publishing GmbH
Trappentreustraße 1
80339 München

Print and binding: Books on Demand GmbH, Norderstedt, Germany
Printed on acid-free paper from responsible sources.

The present work has been carefully prepared. Nevertheless, authors and publishers do not incur liability for the correctness of information, notes, links and advice as well as any printing errors.

GRIN web shop: https://www.grin.com/document/1366996

CLARIAS GARIEPINUS FISH AS A SUITABLE CULTURE FOR CARRION INSECTS IN FORENSIC INVESTIGATIONS USING DICHLORVOS AS A SUICIDE AGENT

BY

EMMANUEL TYOKUMBUR

1

DEDICATION

We dedicate this scientific study to God Almighty.

ABSTRACT

A study was carried out on the forensic entomotoxicological evaluation of carrion insects of fishes poisoned with dichlorvos on the main campus of the University of Ibadan. Fishes (*Clarias gariepinus*) for the study were obtained from the fish farm on the University Campus. The fishes were then euthanized with 5ml of dichlorvos while the control was without poison. Adult carrion insects were collected from the fish carrion using a sweep net. The larvae were collected into a bowl by using a scoop, immobilized with hot water and later placed in sample bottles containing 70% ethanol. Pupae were collected using forceps. Carcass temperature was measured using infrared thermometer while relative humidity was recorded from a digital hygrometer. Calliphoridae and Muscidae were the initial pioneers of the decomposing carcass and were seen during the fresh stage, sarcophagidae was seen shortly after the fresh stage of decomposition. The highest mean temperature value for the fish treated with 5ml Dichlorvos was 31.3^0C while its lowest mean value was 22.6^0C. It was observed that the temperature on the ninth day was higher than the other days. This is attributed to the heat generated by the active maggots at that stage of decomposition. Dichlorvos was found to retard the growth of carrion larva as shown in *Musca domestica* larvae when compared with the control. Fast decomposition rate was recorded due to high ambient and carcass temperature. It can be concluded from this study that carrion insects can be used in solving crime puzzles through the extrication of post mortem intervals in conjunction with environmental variables. Since fish carrion in this study attracted a sizable number of carrion insects, it is recommended that fish be used in future forensic case and experimental studies.

Keywords: Entomotoxicology, Dichlorvos, Clarias gariepinus, Carrion insect, Decomposition

TABLE OF CONTENT

LIST OF TABLES

LIST OF FIGURES

6

CHAPTER ONE

1.1   Entomotoxicology

Entomotoxicology is derived from "entomo" which literally means insects and "toxicology" which means the study of toxins. Therefore entomotoxicology can simply be defined as the study of toxins or poisons in insects (Amendt et al, 2004).

Forensic entomology is the scientific study of the invasion and succession pattern of arthropods with their developmental stages of different species found on the decomposed cadaver during legal investigations. It is the application and study of insect and other arthropods, including insects, arachnids and centipedes to criminal or legal cases. It is primarily associated with investigations (Amendt et al, 2004)

Forensic entomotoxicology is most commonly used to estimate Post Mortem Interval (PMI) in cases involving homicide (Wallman,2003: Dadour and Morris,2009). The initial signs of soft tissue decay occurs within the first 72-98 hours after death. After the initial stages of decomposition are complete, the accurate determination of PMI is not possible but insects found on the deceased can enable entomologist to provide an estimation ranging from one day up to more than 2 months (Schoenly et al, 1992).

1.2   Carrion Insects

Carrion insects are insects that are associated with decomposing or deteriorating remains (Nuorteva, 1977). Carrion insects are very important in the ecosystem as they help in the breakdown of dead and decomposing organisms and they also return nutrients to the soil.

Insects are attracted to a body immediately after death, and they colonize in a predictable manner. A corpse, whether human or animal, is a large food resource for a great many creatures, and supports a large and rapidly changing ecosystem as it decomposes. The body progresses through a recognized sequence of decompositional changes, from fresh to skeletal, over time (Nuorteva, 1997). During this decomposition, it goes through dramatic, physical, biological, and chemical changes (Henssge et al., 1955; Vaqn den Oever 1976; Coe and Curran 1980). Each of these stages of decomposition is attractive to a different group of Sarcosaprohagous arthropods, primarily

7

insects. Some are attracted directly by the corpse, which is used as food or an oviposition medium, whereas other species are attracted by the large aggregation of other insects they use as food resource.(Henssge *et al.*, 1955; Vaqn den Oever 1976; Coe and Curran 1980)

1.3 Dichlorvos

Dichlorvos (2,2-dichlorovinyl dimethyl phosphate) commonly abbreviated as an DDVP is an organophosphate widely used as an insecticide to control household pests, in public health, and protecting stored from insects. Dichlorvos, like other organophosphate insecticide, acts on acetylcholinesterase, associated with the nervous system of insects. Evidence for other modes of action, applicable to higher animals, have been presented. It is claimed to damage DNA of insects (Haz-map, 2015)

Dichlorvos is widely used as a suicide agent in developing countries. The oral LD_{50} of dichlorvos in its various forms for rats ranges from 25 to 80mg/kg, 61 to 175mg/kg in mice, 100 to 109mg/kg in dogs, 15mg/kg in chickens, 157mg/kg in pigs, and 11 to 12.5mg/kg in rabbits. The dermal LD_{50} for dichlorvos in rats is 70.4 to 250mg/kg, 206mg/kg in mice, and 107mg/kg in rabbits. The 96-hour LC_{50} for dichlorvos in fathead minnow is 11.6mg/L, 0.9mg/L in blue gill, 5.3mg/L in mosquito fish. The 24-hour LC_{50} for dichlorvos in blue gill sunfish is 1.0mg/L (Howard, P. 2017)

1.4 Aims and Objectives

i   To assess the effect of Sniper insecticide on the developmental stages of insect larvae found   on *Clarias gariepinus*.

ii   To assess the abundance and species diversity of carrion insects found on *Clarias gariepinus*.

iii To study the carcass colonization and succession pattern of the carrion insects found on the   decomposing fish.

iv   To evaluate the bioaccumulation of dichlorvos in the insect larva found on the carcass of the   poisoned *Clarias gariepinus*

CHAPTER TWO

LITERATURE REVIEW

2.0. INTRODUCTION

Forensic entomology is the study of application of insect and other arthropods in criminal investigation (Joseph *et al.*, 2011). Insect or arthropods are found in decomposing vertebrate corpse or carrion. These insect colonizers can be used to estimate the post mortem index (PMI) movement of the corpse, manner and cause of death and association of suspects at the death scene.

There are three main fields in forensic entomology, and they are:

1. Urban Forensic Entomology
2. Stored-product Forensic Entomology
3. Medico-legal Forensic Entomology

2.1   URBAN FORENSIC ENTOMOLOGY

This field of forensic entomology focuses on pest infestations that are related to litigation such as legal disputes between exterminators and landlords. Besides, studying insects and other arthropods, urban forensic entomology typically also involves toxicology studies, for example the appropriateness of pesticide application. (Catts and Goff, 1992)

2.2   STORED-PRODUCT FORENSIC ENTOMOLOGY

This field of forensic entomology investigate cases of insect infestation or contamination of commercial food with an aim to find evidence related to litigation (Catts and Goff, 1992).

2.3   MEDICO-LEGAL FORENSIC ENTOMOLOGY

This sub-field of forensic entomology gather evidence through the use of insects and other arthropods at a crime scene, such as murder and suicide. It most often involve the study of insect eggs and maggots in what order did they appear and where on the body they are found. Because insects occur only in particular places and are active only at a particular season, their presence can reveal a lot about location and time of the crime. Arthropods and their association with post modem changes of the human body. As seen as death occurs, cell starts dying and enzyme start digesting the cells inside out in a process called autolysis. The body starts decomposing (Catts and Goff, 1992).

Based on the studies done by Craig *et al.*,(1950), it was found that putrative sulphur-based compounds were responsible for initially attracting the flies to the decomposing carcass but egg laying or oviposition of the flies are induced by ammonium-rich compounds present on the carrion.

2.4   Decomposition

Decomposition is the process of by which organic substances are broken down into simple organic matter. The process is a part of the nutrient cycle and is essential for recycling the finite matter that occupies physical space in the biosphere (Keh, 1982)

The decomposing activity of bacteria, fungi and carrion animal feedings is the main process by which nutrients are released and recycled from dead animals. Arthropods are mainly carrion feeding vectors represent the most conspicuous element in this process (Nuorteva, 1977; Hanski, 1987). The role of necrophagous flies is ecologically important given their predominance community assemblages are very complex and highly competitive usually dominated by one to three species (Hanski, 1987; Ives, 1991)

2.4.1   Stages of decomposition

2.4.1.1. Stage 1: The living fish

A live fish is not outwardly decomposing, but its intestine contains a diversity of bacteria, protozoan and nematodes. Some of these micro-organisms are ready for a new life, should the fish die and lose its ability to keep them under control (Smith, 1986).

2.4.1.2   Stage 2: Initial decay

Although the body shortly after death appears fresh from the outside, the bacteria that before death were feeding on the contents of the intestine begin to digest the intestine itself. They eventually break out of the intestine and start digesting the surrounding internal organs. The body's own digestive enzymes (normally in the intestine) also spread through the body, contributing to its decomposition. On an even smaller scale, enzymes inside individual cells are released when the cell dies. These enzymes break down the cell and it's connections with other cells (Tantawi *et al.*, 1996)

From the moment of death, flies are attracted to bodies. Without the normal defences of a living animal, blow flies and house flies are able to lay eggs around wounds and natural body openings (mouth, nose, eyes, anus, genitalia). These eggs hatch and move into the body, often within 24 hours. The life cycle of a fly from egg to maggot to fly takes from two to three weeks. It can take considerably longer at low temperatures (Smith, 1986).

2.4.1.3. Stage 3: Putrefaction

Bacteria break down tissues and cells, releasing fluids into body cavities. They often respire in the absence of oxygen (anaerobically) and produce various gases including hydrogen sulphide, methane, cadaverine and putrescine as by-products. People might find these gases foul smelling, but they are very attractive to a variety of insects (Catts and Goff, 1992).

The build up gas resulting from the intense activity of the multiplying bacteria, creates pressure within the body. This pressure inflates the body and forces fluids out of cells and blood vessels and into the body cavity (Catts and Goff, 1992).

The young maggots move throughout the body, spreading bacteria, secreting digestive enzymes and tearing tissues with their mouth hooks. They move as a maggot mass benefiting from communal heat and shared digestive secretions (Catts and Goff, 1992).

2.4.1.4. Stage 4: Black putrefaction

The bloated body eventually collapses, leaving a flattened body whose flesh has a creamy consistency. The exposed parts of the body are black in color and there is a very strong smell of decay (Smith, 1986). A large volume of body fluids drain from the body at this stage and deep into the surrounding soil. Other insects and mites feed on this material. The insects consume the bulk of the flesh and the body temperature increases with their activity. Bacterial decay is still very important, and bacteria will eventually consume the body if insects are excluded(Catts and Goff, 1992).

By this stage, several generations of maggots are present on the body and some have fully grown. They migrate from the body and bury themselves in the soil where they become pupae. Predatory maggots are much more abundant in this stage, and the pioneer flies cease to be attracted to the corpse. (Smith, 1986).

2.4.1.5   Stage 5: Butyric fermentation

State of decay

All the remaining flesh is removed over this period and the body dries out. It has a cheesy smell, caused by butyric acid, and the smell attracts a new suite of corpse organisms. The surface of the body that is in contact with the ground becomes covered with mound as the body ferments (Catts and Goff, 1992)

The reduction in soft food makes the body less palatable to the mouth-hooks of maggots, and more suitable for the chewing mouthparts of beetles. Beetles feed on the skin and ligaments. Many of these beetles are larvae. They hatch from eggs, laid by adults, which fed on the body in earlier stages of decay (Smith, 1986).

2.4.1.6   Stage 6: Dry decay

The body is now dry and decays slowly. Eventually all the flesh disappears leaving the bones only (Bourel et al., 1999)

Animals which can feed on hair and flesh include tineid moths, and micro-organisms like bacteria. Mites, in turn, feed on these micro-organisms (Bourel et al., 1999). They remain on the body as long as traces of flesh remain, which depends on the amount of flesh that covers the particular species (Bourel et al., 1999)

2.5   Biology of Carrion insects

Carrion-breeding insects have substantial ecological and forensic importance. Fly development is also influenced by the biology of the necrophagous insects. Appetite and reproductive state of adult blow flies can influence whether they are ready to use carrion as a food source of medium for oviposition (Dethier and Bodenstein, 1958)

Flies are attracted to a body immediately after death (Anderson and Vanlaerhoven, 1996; Erzinclioglu, 1983; Nuorteva, 1977). The first flies to be attracted to a body are usually the blow flies, the large metallic flies seen near food or garbage cans.

The blow flies develop from egg through first, second and third instar stages, and pupal stage before becoming adults following a predictable pattern that is primarily influenced by species and temperature (Anderson and Cervenka, 2002)

The sequence of insects that colonize depends on the nutritional changes in the body and is greatly impacted by geographic region, habitat, season, meteorological conditions and microclimate, but the sequence is predictable within these parameters (Anderson, 2001). This predictable and sequential colonization of the body allows an entomologist to determine the tenure of the insects on the body, and the minimum time since death. Changes in climatic factors mainly temperature, relative humidity and rainfall affect carrion decomposition rates (De Carvalho and Linhares, 2001) and subsequently the number of days to complete decomposition.

2.5.1.   Types of Carrion Insect

2.5.1.1   Calliphoridae

Out of all insects visiting a dead body, the maggots of blow flies (Calliphoridae) and to lesser extent flesh flies (Sarcophagidae) are responsible for the maximum consumption of terrestrial carrion (Fuller, 1934; Payne, 1965; Putman, 1977 and 1987; Braack, 1981; Early and Goff, 1986). Blowflies lay eggs on the dead body and the emerging larvae quickly invade most of the regions of the carcass(Putman, 1978).

The major influences that affect blow fly establishment and larval development on carrion are competition, carrion size, temperature and humidity (Denno amd Cothran, 1976; Introna *et al* 1991). Calliphorids commonly exploit the food resource first, thereby out-competing other fly species.

2.5.1.2 Sarcophagidae

The sarcophagids are called flesh flies. Denno and Cothran (1976) found that the life cycle of sarcophagids was generally shorter than that of calliphorids. The feeding larvae of sarcophagids develops more rapidly and in the fall, they pupate and enter diapauses.

Sarcophagids are easily identified into family in the field, but the examination of male genitalia (hypopygia) is the only method for identification to the genus or species level, with exception of one species (Aspoas,1994). Females cannot be identified unless they are gravid and the eggs are reared to produce some adult males for dissection.

2.5.1.3 Muscidae

Muscids are very common flies in a variety of habitats and some are strongly associated with human environments. Species in this family are found breeding and feeding on refuse, excrement and carrion (Carvalho *et al*, 2001. Because of their food preference, muscids are more commonly associated with carrion during the decay and post-decay stages. The forensic importance of this family has not been well documented. Many researchers record their presence and/or abundance, but find this to be too variable to be used as an indicator for time of death.

2.5.1.4 Piophilidae

Piophilids are scavengers and some are considered serious pests of cheese and preserved meats (Borror *et al*, 1989). Many larvae also feed and develop in excrement and carrion. Piophilacasei is commonly known as the cheese skipper and is named for the jumping movement of the larvae. This characteristic movement makes it easy to recognize these larvae in the field. The larvae feed and develop in the thick pastry areas of the carcass. Since larval piophilids carry out their

14

development during this later stage of decomposition and are the dominant species at this time, they are used as an indicator of time of death.

2.5.1.5 Nitidulidae

They are known as the sap beetles. Most species of sap bottles are attracted to the wounds of trees where they feed on sap. They have been found in various habitats feeding on flowers, fruits, sap, fungi, decaying and fermenting plant tissues or dead animal tissue (Carvalho *et al*, 2001)

Sap beetles are characterized by a rather short larval development and comparatively long lived adults. This allows the sap beetles to master and to adapt to extremely different types of substrates.

Sap beetles have demonstrated a wide range of feeding habits with the majority of the group being primarily saprophagous and mycetophagous (Carvalho *et al*, 2001)

2.5.1.6 Silphidae

Most silphids are carrion feeders (necrophagous species) but can also prey on other carrion inhabitants such as fly eggs or maggots and other small carrion beetles (necrophilus species)(Racliffe, 1996; Hastir *et al*, 2001; Sikes, 2005; Sikes, 2008).

They promote the breakdown and recycling of organic matter into terrestrial ecosystems (Racliffe, 1996; Hastir *et al*, 2001; Kalinova et al, 2009). There are some species of silphidae that feed on soil invertebrates (snails, caterpillars or slug predators)

In carrion beetles community, niche differentiation can occur along dimensions of season, habitat and carcass size (Scott, 1998; Hocking *et al.*, 2007)

2.5.2. The Utility of Carrion Beetles in Forensic Entomology

Most forensic researchers have focused on flies while beetles have been neglected (Midgley *et al*, 2009; Midgley *et al.*, 2010). When a corpse colonized by insects is found, two situations could be considered (Amendt *et al.*, 2007; Lefebvre *et al*, 2009). In the first situation, which is the most frequent case in forensic investigations, insects are pioneer species and the minimum post mortem interval (PMI) is estimated with the age of the oldest specimens found on the death scene, principally blow flies (Amendt *et al.*, 2007; Lefebvre *et al*, 2009). In the second situation, later necrophagous species colonize the corpse with a delay, often after the departure of pioneer species. The estimation of the PMI is only possible by analyzing the chronological succession (Amendt *et al*, 2007; Lefebvre *et al*, 2009)

A frequent objection against the use of coleopteran in forensic investigations is the fact that flies (pioneer species) locate corpses faster than beetles (later necrophagous species). Thus, the minimum post mortem interval estimates are more accurate with diptera, especially with the families of calliphoridae and sarcophagidae (Smith, 1986; Midgley *et al*, 2010). However, recent researches have shown that some silphidae can locate a corpse within 24h and their larvae have been observed soon after death, during the early stage of decomposition (Midgley *et al*, 2009; Midgley *et al*, 2010). This implies that some carrion beetles have the same forensic interesting characteristics than carrion flies and can be considered as pioneer species. In this case, some species of coleopteran can be used as blow flies (Midgley *et al*, 2009; Midgley *et al*, 2010)

2.6. Effects of Temperature and Humidity

Bodies found in direct sunlight will be warmer, heating up more rapidly and decomposing faster. They will loose biomass more rapidly than bodies in shade and progress through the decompositional stages faster (Dillion, 1997; Dillion and Anderon, 1958; Shean *et al*, 1993)

Ideally, insects should be collected directly from the body at the death scene. In many questionable death situations, collection of entomological evidence is left until after the body has been removed and chilled at the morgue (Johl and Anderson, 1996)

2.7. Review of Previous Literatures

The pioneer study in modern forensic entomology was carried out by the French doctor Bergeret who was able to detect post-mortem interval (PMI) (Benecke, 2001).in assessing the PMI, he assumed that metamorphosis involves one year in *Musca carnaria L* and that the females lay eggs in summer so that the larvae would transform to pupae the next spring and hatch in summer. He found the eggs of *Musca carnaria L* on the corpse that lays eggs before the body dries out. Using these findings he calculated that the body must have been left there at least a couple of years back. This shows that in spite of limited knowledge and resources, forensic entomology was handy in solving a crime puzzle at that time.

Entomological investigations are common outside the United States. A few of the earliest works in Europe include those by Redi (1668), Bergeret (1855), Brouardel (1879), Megnin (1888, 1894), and Motter (1898). Additional case studies and experiments from the twentieth century include those conducted in such diverse countries as Finland (Nuorteva *et al*, 1967), Italy (Introna *et al*, 1998), Australia (Palmer 1980), South Korea (Rueda *et al,* 1997), Egypt (Tantawi *et al*, 1996 and 1997), and Japan (Nishida 1984).

One of the earliest works completed by Redi (1668) consisted of leaving meat exposed to, or unprotected from flies. This work showed that spontaneous generation of flies did not occur, and that flies contributed to the decay process. Bergeret (1855) later did medicolegal forensic entomology work which provided insight into similar methods and materials still used in today's investigations. But it was Megnin's work from 1883-1898 which suggested for the first time that the process of decomposition occurred in a predictable manner. These early studies gave status to the budding branch of entomology among the forensic sciences.

Goff *et al*, 1994 have examined the evidence of the effects of cocaine, heroine, and other various chemicals upon decaying tissues and the arthropods associated with them. Goff and Lord (1994) refer to this new field as entomotoxicology and believe it has good prospects of becoming a standard forensic technique when insects and chemicals are found together.

Campobasso *et al* (2004) conducted a study of the correlation level of accumulation of various toxic substances in human tissue and the larvae of *Lucilia sericata* (Meigen, 1826) feeding on that tissue. The tissue initially underwent toxicological analysis and was subsequently examined for the concentrations of chemical compounds detectable in the larvae of true flies feeding on it. The analysis demonstrated that only in the case of cocaine, the concentration was similar in both tissues (human and larval), while other substances were found at much lower concentrations in the larval tissues than in the control tissue.

The rate of decomposition changes according to the temperature (the most important variable in all developmental models), environment (soil or vegetation in an open-air setting or conditions in an enclosed space), weather, as well as corpse condition and exposure (exposed or buried, naked or dressed, intact or wounded) that occur at the site of discovery of the body (Gennard, 2007; Matuszewski *et al* 2016).

Miller *et al* (1994) first reported the first detection of drugs from chitinized insect tissues that also include the exuviae. The study was based on the development of hair extraction technologies, in which attention has recently focused on the analysis of chitinized insect remnants which are frequently encountered with mummified or skeletonised remains. They emphasized that the detection of various toxins and controlled substances in insects found on decomposing bodies can contribute to the assessment of cause or manner of death and recommend the use of anthropophagic fly larvae (maggots) as alternative toxicological specimens.

Musvasva *et al* (2001) studied preliminary observations on the effects of hydrocortisone and sodium methohexital on development of *Sarcophaga* (Curranea) tibialis Macquart (Diptera: Sarcophagidae), and implications for estimating post mortem interval. They found no systemic relationship between drug concentration and development time of larva and pupae which imply that emergence and post-mortem interval (PMI) will not be affected by these drugs.

In Nigeria, forensic science in general has not been fully in application in the legal system especially criminal cases relating to homicides and questionable deaths. This had however negatively affected other branches of forensic science, including forensic entomology. Hence, the report of (Usua, 2007) that forensic entomology in Nigeria is still in its infancy stage, was a clarion call which has stimulated interests in forensic entomology. Few entomologists have responded to the call thus, studies are gradually spreading from the Southern Nigeria to the Eastern Nigeria, in the last few years. These studies were at least to provide entomological base line information on the dearth knowledge of our legal system, that insects can provide useful answers to questionable deaths.

To add to the baseline information for our legal system, efforts were made to study the insect fauna associated with the decomposing pig carrions as models, to enhance unbiased delivery of justice in the case of homicides and questionable deaths in Nigeria. The study critically observed different stages of decomposition of the pig carrions and marking the stages of the decomposition with time that had elapsed cum insects that are distinctively unique on each decomposition stage.

The carrion feeding insects (flies and beetles) provide important ecological services on dead animals. In case of suspicious death, forensic entomologist can use insect evidence to help forensic investigators determine what had happened to the victim and use documented information about the life cycles and guild structures and community dynamics of these insects to determine facts like the time of death. Hence, the insects can be useful in criminal investigation relating to homicides and questionable deaths in the court of law.

CHAPTER THREE

MATERIALS AND METHODS

3.1 Study site

The site to be used for the study is located on the departmental grounds (Department of Zoology) near the Animal House. The site is suitable as there is limited movements and activities around the site. So the smell emanating from the decomposition of the fish carcass will pose little or no threats to humans. The location reading on the digital compass is $7^0 26'37''$N $3^0 53'46''$E.

3.2.1 Preparation of fishes for setup

The fishes were transported from the farm to the field site using a bucket containing water. The bucket was covered with mesh net to prevent the fishes from jumping out. Measuring cylinder was used to measure out 5ml of Dichlorvos and administered to the two fishes through the mouth and left to die. The control fishes (2) were allowed to die without dichlorvos intake. This was achieved by percussive stunning to induce immediate insensibility by administering a severe blow to the skull of the fish. The fish then remains unconscious until death.

3.2.2 Mounting the fishes

The fishes after death were placed on separate mounts. Two fishes per mount. The mounts consisted of a tray covered with sawdust and placed on a stool. The essence of the sawdust is to provide shelter for the developing larvae and pupae of the insects. The stool is important as it provides a platform that excludes other animals that are not of entomological importance that might also visit the carcass. Also the container with spent engine oil placed on each leg of the stool is able to trap other arthropods that are not of entomotoxicological importance that might also visit the carcass. "Keep Off" signs were also placed on each carcass mount to better inform people who might encounter the setup.

3.2.3 Carrion insect collection and processing

The carrions will be collected in the following order:

First, those flying over and/or landing on the carcass; then those found in natural cavities and finally those that are under the carcass and in the sawdust upon which the carcass is placed. Sampling for adult insects will be done by using a sweep net and insecticide. The net will be swept clockwise and anticlockwise at an angle of almost 180^0 arc over the decomposition carcass after which the open end will be quickly folded by using the second hand so as to prevent the escape of the trapped insects. The insecticide will then be sprayed over the sweep net to immobilize the insects and they will be transferred into appropriately labelled clean and clear sample bottles. The collected samples will be fixed in 70% alcohol

3.2.3.1 Sampling for larvae

Maggot is the larvae stage of carrion insects. This involves using a sampling spoon to collect adequate amount of the maggots from the decomposing carcass. The active maggots will then be transferred into a small bowl. Hot water kept in a flask will then be poured into the bowl containing the maggots. The hot water kills and renders the maggots inactive. The maggots will then be isolated by using a sieve and then placed into appropriately labelled sample bottles.

3.2.3.2 Sampling for pupae

When the maggots on the decomposing carcass start reducing, the sawdust was checked for the presence or absence of pupae. If pupa were present, they were collected and also kept in appropriately labelled sample bottles.

3.3 Measurement of larval body length

The lengths and weights were measured and mean values recorded for each carrion group at different stages of decomposition. Mean values were used for statistical analysis. Time of pupation and adult emergence were recorded for each carrion group. This was followed by group mean developmental period. Length of larvae from the second instar stage were obtained by a pair of divider and read on a transparent meter rule.

3.4   Measurement of carcass temperature and humidity

The temperature of the carcass were read and recorded daily using an infrared thermometer that can accurately measure between -50^{0}C to 330^{0}C. The thermometer was used by pointing the infrared beam towards the pig carcass. Readings were then generated and shown on the thermometer screen. The readings were then recorded into the field experiment book. Measurement of relative humidity was done using a digital hygrometer. The hygrometer is placed as close as possible to the carcass, readings are generated on the LCD screen of the hygrometer and the readings are then recorded into the field experiment book.

CHAPTER FOUR

RESULTS

4.1 Abundance and species composition of carrion insects on fish carcass.

The result of the abundance of forensically important insects collected from Clarias treated with 5ml of Dichlorvos is shown in Table 1. From the table, calliphoridae and sarcophagidae leads the fresh stage with 2 representatives each, muscidae had one representative. In the bloat stage, calliphoridae proved to be the most abundant with 8 representatives, sarcophagidae had 4 representatives and muscidae lagged with 2 representatives. At the dry stag of decomposition, muscidae was the most abundant with 6 representatives and a new family; dermastidae came up with 2 representatives.

Family	Genus/species	Fresh	Bloat	Active	Advanced	Dry
Calliphoridae	*Lucilia sericata*	2	4	8	2	0
Sarcophagidae	*Sarcophaga spp*	2	4	4	6	0
Muscidae	*Musca domestica*	1	2	2	2	6
Dermastidae	*Dermestes maculates*	0	0	0	1	2

Table 1: Abundance of forensically important insects collected from fish carcass treated with 5ml of Dichlorvos

Table 2 shows the abundance of forensically important insect collected from fish carrion in the control group. Calliphoridae is the abundant family with 4 representatives, the muscidae follows with 2 representatives. In the bloat stage, family calliphoridae was still found to be dominant. The calliphoridae was found dominating all stages except advanced and dry stages. The family dermestidae was found dominating all stages except advanced and dry stages. The family dermestidae was found in the active, advanced and dry stages with representatives in the format 1, 1, 2 respectively.

Family	Genus/species	Fresh	Bloat	Active	Advanced	Dry
Calliphoridae	*Lucilia sericata*	4	8	8	0	0
Sarcophagidae	*Sarcophaga spp*	1	1	4	1	0
Muscidae	*Musca domestica*	2	4	1	0	0
Dermastidae	*Dermastes maculates*	0	0	1	1	2

Table 2: Abundance of forensically important insects collected from control fish carcass

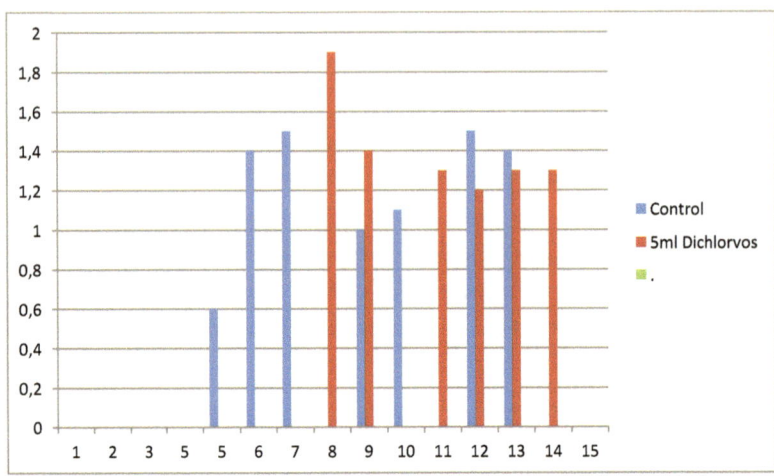

Fig. 1: Histogram showing the mean body length of the larvae of *Musca domestica* collected from fish carcass

Figure 1 shows the effect of Dichlorvos on the mean body length of the larvae stage of *Musca domestica* collected from the fish carcass. We can observe that the first four days after death of fish was required for transformation of *Musca domestica* into the second instar stage. The second instar larvas were collected from the decomposing fishes on the fifth day.

On the sixth day, the mean length of larvae from the control group was 1.4cm increasing by 0.8cm from the mean larva length of the fifth day.

On the seventh day, the mean length of the larvae from the control was increased by 0.1cm. on the eighth day, the mean length of the larvae in the poisoned carcass shot up to 1.9cm and no mean larva length was recorded for the control group.

On the ninth day, the mean larvae length for the control group was 1.0cm and the mean larvae length for the poisoned fish was 1.4cm

There was no record for the poisoned group on the tenth day but the mean larvae length for the control group was placed at 1.1cm

On the eleventh day, there is a record for the mean larvae length for the poisoned fish and was found to be 1.3cm. The length reduced in comparison to the mean larvae length for the poisoned fish on the eight day.

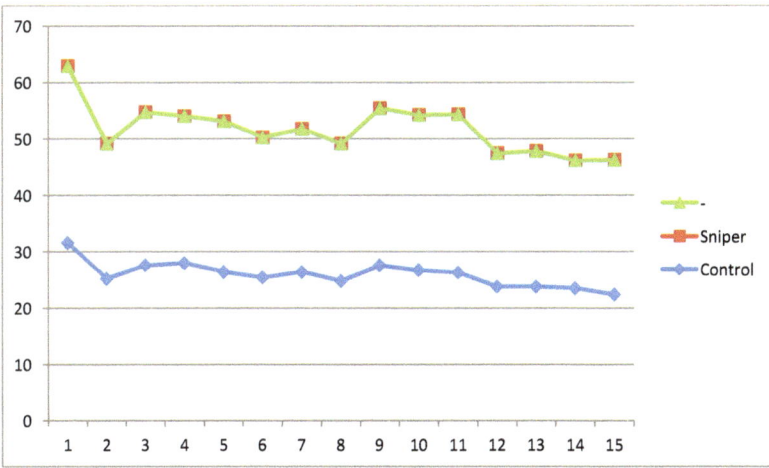

Fig 2: Comparison of Dichlorvos poisoned and control fish carcass temperature recorded in the field experiment

From the graph in Fig 2. We notice that the mean temperature of the control group ranges from 31.6^0C which was the highest value recorded to 22.4^0C which was the lowest value recorded. The highest mean temperature value for the fish treated with 5ml Dichlorvos was 31.3^0C while its lowest mean value was 22.6^0C. It was observed that the temperature from day 9-11 was higher than the other days. This is attributed to the heat generated by the active maggots at that stage of decomposition.

PICTURES OF CARCASS DECOMPOSITION STAGES IN THE STUDY

Fig 3: Fresh decomposition stage

Fig 4: Bloat decomposition stage

Fig 5: Active decomposition stage

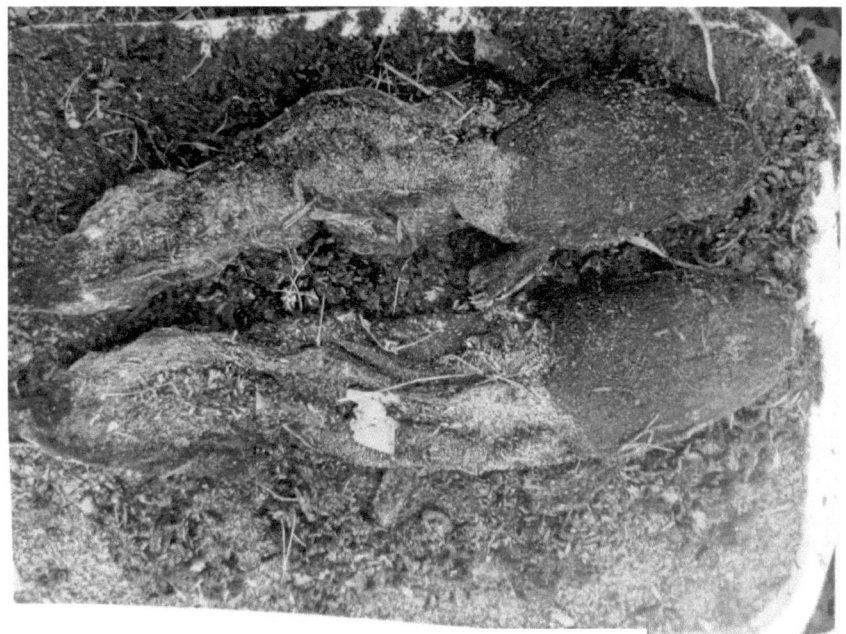

Fig 6: Advanced decomposition stage

Figure 7: Skeletonized remains of fish carcass

CHAPTER FIVE

DISCUSSION

5.1   Abundance and species composition of carrion insects on fish carcass

At the end of the field work, five stages of decomposition were observed and this corroborates earlier findings and experiments such as the one done by Carvalho *et al*., (2004) where they found out that the five important stages of decomposition are the fresh stage, bloated stage, active stage, active stage, advanced stage and the dry stage. The four major families of carrion insects in the study were also reported in Carvalho *et al*., (2004) experiment, although Carvalho *et al* recorded some families that were not observed in this study such as Formicidae, Histeridae, Staphylinidae, etc. especially in the post decay stage. Abajue *et al*., in 2013 also recorded that the arthropods that arrived on a carcass from the beginning of the decomposition are the Calliphoridae, Saecophagidae and Muscidae.

5.2   Carrion insect succession on fish carcass

During the experiment, Calliphoridae and Muscidae were the initial pioneers of the decomposing carcass and were seen at the fresh stage, while Sarcophagidae arrived shortly after the fresh stage of decomposition. Dermastidae was later seen in the advanced stage of decomposition and it was observed till dry stage. The trend we observed here was similar to the trends observed by Abajue *et al*., (2013) and Ekrakene and Iloba (2011)

5.3   Effects of Dichlorvos on the length and weight of larva

In this study, dichlorvos was found to inhibit the growth of *Musca domestica* and when compared with the control this corroborates earlier findings by Ekrakene and Odo (2017) when they assessed effects of varying volumes of cypermethrin pesticide on the larval body length, weight, and development time of blowfly *Chrysomya albiceps* (Diptera: Calliphoridae) reared on rabbit carrions.   According to an experiment by Andrew (2014), where he studied the abundance and occurrence of carrion insect on pigs poisoned with nicotine, he discovered that the low group had lesser occurrence and abundance of carrion insects when compared with the control group, he also discovered that the high dose group nicotine poisoned pigs had more occurrence and abundance of carrion insects than both the low dose group and control group. However, in this study, dichlorvos was found to inhibit larva growth.

5.4   Variation in environmental variables at the field site

The high tropical temperature and relative humidity of Ibadan where the experiment was conducted aided in the faster decomposition of the fishes. The result of the fast decomposition agrees with Ekanem and Dike (2010) where it was established that higher air temperature leads to a faster decomposition rates and increase in the abundance of insects.

5.5   CONCLUSION

Decomposition is a natural process to organic material back into the ecosystem. Corpses are the suitable microhabitat for certain organisms because of its food source and shelter. In the latest years more research has been conducted in the field of entomotoxicology, trying to find a relationship between drug concentrations in the substrate usually a carrion animal and insects found on that carrion, and to increase the knowledge of insect development. At the end of the experiment I was able to establish the fact that indeed suicidal poisons will affect the diversity, abundance and decomposition of carrion insects found on decomposing fishes. As the dosage of Dichlorvos poisoning increases, the composition, abundance and diversity of carrion insects on the decomposing fish will also reduce.

REFERENCES

Abajue, M. C. Ewuim, S.C. and Akunne, C.E. (2013). Insects associated with decomposing pig carrions in Okija, Anambra State, Nigeria. *The Bioscientist,* 1(1), 54-59.

Adams, Z. J. And Halls M.J. (2003). Methods used in the killing and preservation of blowfly larvae and their effects on the PMI Length. *Foren. Sci. Int.,* 138:50-61.

Amendt, J., Krettek, R., and Zehner, R. (2004). Forensic entomology. *Naturwissenenschaften,* 91(2), 51-65.

Anderson, G. S., Cervanka, V.J., Haglund, W., and Sorg, M. (2002). Insects associated with the body: their use and analyses. *Advances in forensic taphonomy: method, theory, and archaeological perspectives, 173200.*

Anderson, G.S. (2001). Insect succession on carrion and its relationship to determining time of death. *Forensic entomology: the utility of arthropods in legal investigations,* 143, 76.

Benecke, M. (2001). A brief history of forensic entomology. *Forensic science international,* 120(1-2), 2-14.

Bharti, M., and Singh, D. (2002). Occurrence of different larval stages of blow flies (Diptera: Calliphoridae) on decaying rabbit carcasses. *Journal of Entomological Research,* 26(4), 343-350.

Bodenstein, D.V. (1958). Hunger in the blowfly. Z. *Tierpsychol,* 15, 129-140.

Borror, D.J., Triplehorn, C.A., and Johnson, N. F. (1989). *An introduction to the study of insects* (No. Ed. 6). Saunders college publishing.

Bourel, B., Luck, M. B., Hedouin, V., Cailliez, J. C., Derout, D., and Gosset, D. (1999). Necrophilous insect succession on rabbit carrion in sand dune habitats in northern France. *Journal of Medical Entomology,* 36(4), 420-425.

Campobasso, C.P. Di Vella, G. and INTRONA, F. (2001). "Factors affecting decomposition and Diptera colonization". Forensic Science International. 120: 18-27.

Campobasso, C.P., Gherardi, M., Caligra, M., Sironi, L., and Introna, F. (2004). Drug analysis in blowfly larvae and in human tissues: a comparative study. *International journal of legal medicine, 118*(4), 210-214.

Carvalho. (2010). Toxicology and forensic entomology. In Amendt J, Campobasso CP, Goff ML, Grassberger M. editors. *Current concepts in forensic entomology.* Dordrecht: Springer; p. 16378.

Catts, E.P., and Goff, M.L. (1992). Forensic entomology in criminal investigations. *Annual review of Entomology, 37*(1), 253-272.

de Carvalho, L. M. L., and Linhares, A. X. (2001). Seasonality of insect succession and pig carcass decomposition in a natural forest area in southeastern Brazil. *Journal of Forensic Science, 46*(3), 604-608.

Dekeirsschieter, J., Verheggen, F. J., Haubruge, E., and Brostaux, Y. (2011). Carrion beetles visiting pig carcasses during early spring in urban, forest and agricultural biotopes of Western Europe. *Journal of Insect Science*, 11(1), 73.

Fratczak, K., and Matuszewski, S. (2016). Classification of forensically-relevant larvae according to instar in a closely related species of carrion beetles (Coleoptera: Silphidae: Silphinae). *Forensic Science, Medicine, and pathology, 12(2),* 193-197.

Greenberg, B. (1985). Forensic entomology: case studies. *American entomologist, 31*(4), 25-28.

Hanski, I. (1987). Carrion fly community dynamics: patchiness, seasonality and coexistence. *Ecological Entomology*, 12(3), 257-266.

Hocking, M. D., Darimont, C. T., Christie, K. S., and Reimchen, T. E. (2007). Niche variation in burying beetles (Nicrophorus spp.) associated with marine and terrestrial carrion. *Canadian Journal of Zoology, 85*(3), 437-442.

Howard, P. (2017). *Handbook of environmental fate and exposure data: for organic chemicals, olume III pesticides. Routledge*

Joseph, I., Mathew, D.G., Sathyan, P., Vargheese, G. (2011). The use of insects in forensic investigations: An overview on the scope of forensic entomology. *Journal of forensic dental sciences,* 3(2), 89.

Keh, B. (1985). Scope and applications of forensic entomology. *Annual review of entomology,* 30(1), 137-154.

Lefebvre, F., and Gaudry, E. (2009, January). Forensic entomology: a new hypothesis for the chronological succession pattern of necrophagous insect on human corpses. In *Annales de la Societe entomologique de France* (Vol. 45, No. 3, pp. 377-392). Taylor & Francis Group.

Miller, M.L., Lord, W. D., Goff, M. L., Donelly, B., McDonough, E. T., and Alexis, J.C. (1994). Isoloation of amitriptyline and nortriptyline from fly puparia (Phoridae) and beetle exuviae (Dermastidae) associated with mummified human remains. *Journal of Forensic Science,* 39(5), 1305-1313.

Myers, L. (2004). *Sap beetles of Florida, Nitulidae (Insecta: Coleoptera: Nitidulidae).* University of Florida Cooperative Extension service, Institute of Food and Agricultural Sciences, EDIS.

Nuorteva, P. (1977). "Sarcosaprophagous insects as forensic indicators". In C.G. Tedeschi; W.G. Eckert; L.G. Tedeschi. *Forensic medicine: a study in Trauma and Environmental Hazards.* II. New York: W.B. Saunders. pp. 1072-1095

Putman, R. J. (1978). Patterns of carbon dioxide evolution from decaying carrion decomposition of small mammal carrion in temperate systems 1. *Oikos,* 47-57.

Schoenly, K. (1992). A statistical analysis of successional patterns in carrion-arthropod assemblages: implications for forensic entomology and determination of the post-mortem interval. *Journal of forensic science,* 37(6), 1489-1513.

Smith, K.G.V. (1986). A manual of forensic entomology. British Museum (Natural History & Cornell University Press)

Tantawi, T. I., El-Kady, E. M., Greenberg, B., and Ghaffar, H. A. (1996). Arthropod succession on exposed rabbit carrion in Alexandria, Egypt. *Journal of Medical Entomology,* 33(4), 566-580.

Von Zuben, C. J., Bassanezi, R. C., Dos Reis, S. F., Godoy, W. A. C., and Von Zuben, F. J.(1996). Theoretical approaches to forensic entomology: I. Mathematical model of postfeeding larval dispersal. *Journal of Applied Entomology,* 120(1-5), 379-382.

Wells, J.D., Stevens, J. R., Byrd, J.H., and Castner, J.L. (2009). Molecular methods for forensic entomology. *JH Castner and JL Byrd (eds). Forensic Entomology: the utility of Arthropods in legal investigations, 2,* 437-452.